Many thanks to you mom and dad for everything you did for me and my sisters.

Scuffleheads Introduce
Sam Thomas and Sarah Tyler

Published by
Scuffleheads Inc.

Character creation: Gunnar Oddsson
Character design: Matt Allen and Gunnar Oddsson
Illustration: Andrea Montano and Gunnar Oddsson

Printed in the United States of America

ISBN
No 978-0-578-17750-2

TM and Copyright 2015 by Scuffleheads Inc.

www.scuffleheads.com

COMMUNICATION

The act or process of using words, sounds, signs, or behaviors to express or exchange information or to express your ideas, thoughts and feelings to someone else.

TELECOMMUNICATION

the act or process of using electronic devices, such as a telephone, cell phone, computer (and apps), etc., to express or exchange information or to express your ideas, thoughts and feelings to someone else.
(mostly done via fiber optics, satellites, and microwaves).

OUR MISSION

With fun, colorful cartoon characters,
we are going to walk you through the wonders of
telecommunication as we know it today.

Our characters will explain to you,
in a very simple way, how you can use the
Internet to get information, watch movies,
send pictures, and much, much more.

What is the source of all this information
and all the movies?

How does your selfie go from your phone
to your friend's phone?

In this first book of many, our friends Sam and
Sarah are going to introduce to you
the most important component that allows you to
use the Internet the way you do today.

Hello, I'm Sam Thomas.

And I'm Sarah Tyler.

You are probably thinking to your-self, "Who are these guys?"

We are called **fiber optic connectors** and we are part of a humongous **fiber optic network** that connects most every computer, TV, and cell phone in the world.

You see, everything you watch on TV or the computer and all the music you download on the computer or your cell phone goes through multiple **fiber-optic networks**. Even the **selfie** you send to your friends goes through a **fiber-optic network**.

Seattle,
Washington

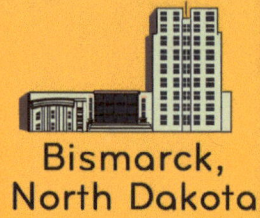

Bismarck,
North Dakota

San Francisco,
California

Cheyenne,
Wyoming

Denver,
Colorado

Santa Fe,
New Mexico

Los Angeles,
California

Dallas
Texas

So what is a **fiber-optic network?**
A **fiber-optic network** is made up of
fiber-optic cables that are
connected to electronic equipment in
every big city and small town all over
the country and all over
the world.

Chicago, Illinois

New York City, New York

Washington D.C.

Columbia, Missouri

Lynchburg, Virginia

Atlanta, Georgia

Miami, Florida

Every continent is connected with **fiber-optic cables,** and before long, every country in the world will be connected with them. We can get and share information with people all over the world in a matter of seconds.

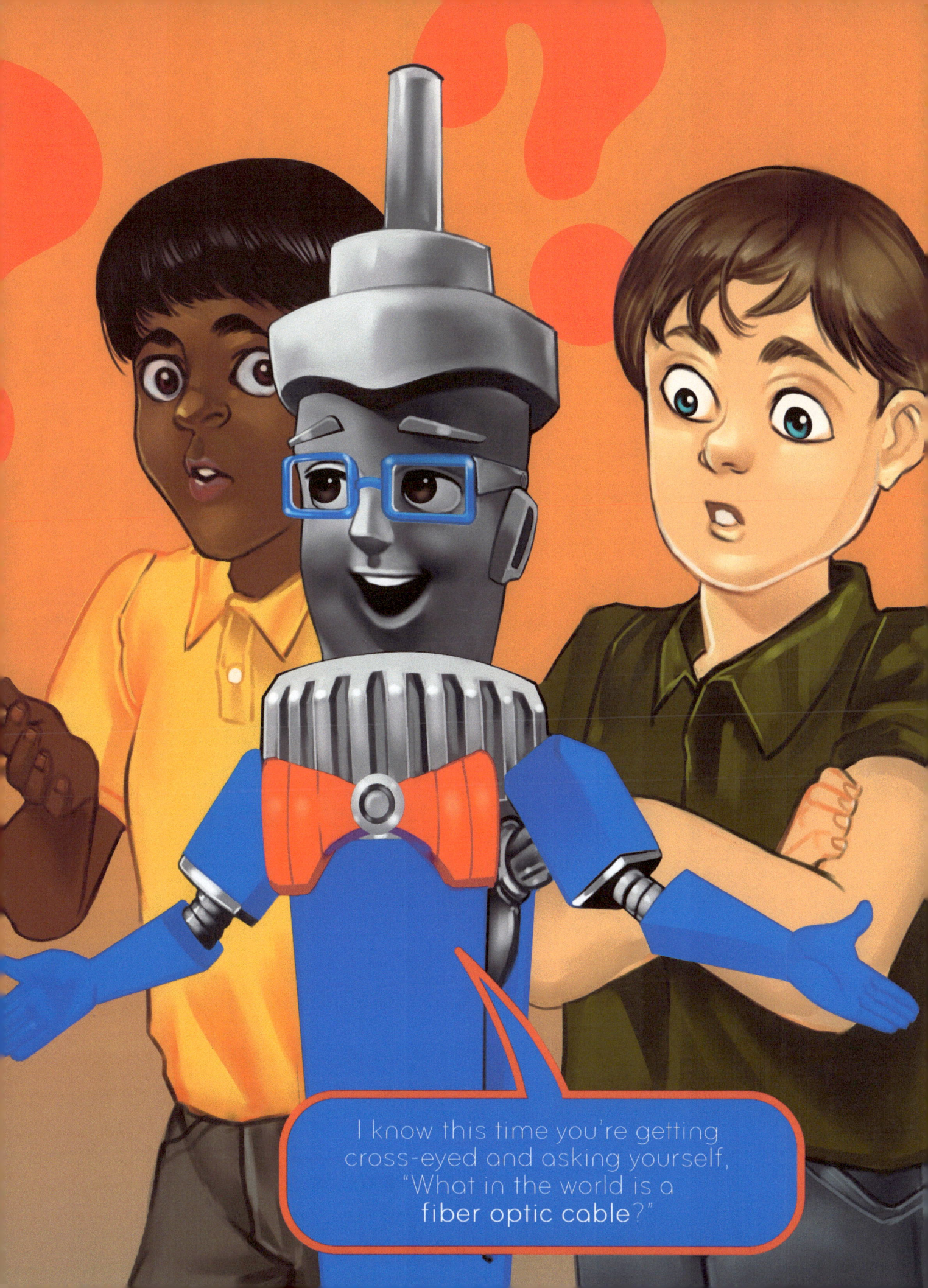

I know this time you're getting cross-eyed and asking yourself, "What in the world is a **fiber optic cable**?"

Fiber-optic cables are made up of
multiple strands of the purest glass
we know, which is called silica.
Each fiber is thinner than a human hair,
and for you to see it better and work
with it, each fiber is coated with
a certain color.

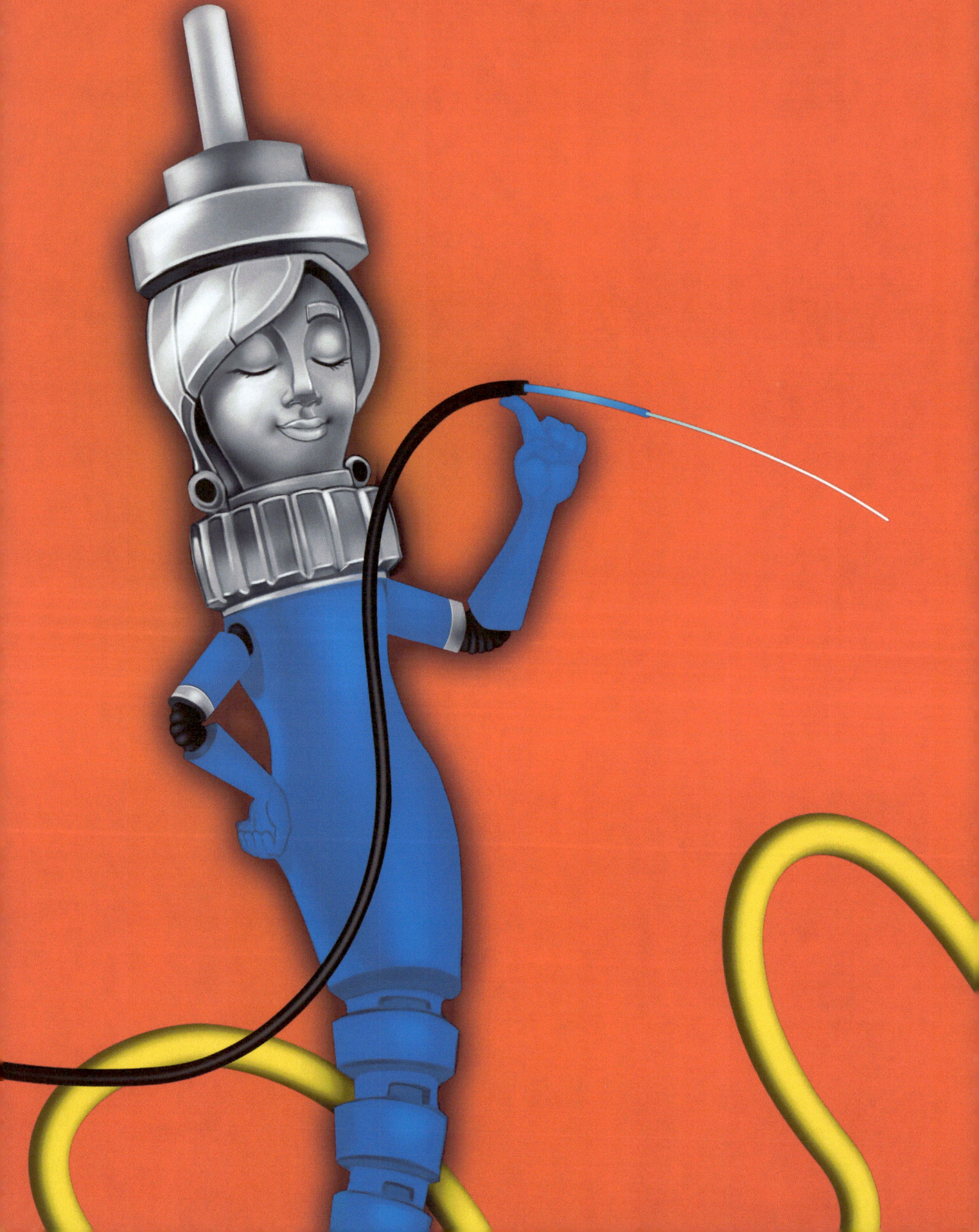

One **fiber** is capable of transmitting many, many more phone calls, images, and other data than the traditional copper wire. With a **fiber** connection, it takes only a few minutes to download your favorite movie. With the old copper wire, it would take almost forever.

When a **fiber-optic cable** is placed in an office building or your house, somebody like Sam or me needs to be put at the end of the **fiber** and connected to the electronic equipment, so that you can watch your favorite show on your TV or your computer.

There are millions of **fiber-optic connectors** like Sarah and me all over the world. But there are also other connectors that are a little different in shape, color, and size.

These guys and others are a part of
our family and friends, and in our next book,
I would like to introduce you to the others
and tell you more about those magnificent
fiber-optic networks that span the globe
and that have brought us
closer together than ever before.

I hope you've enjoyed this short introduction
to what we call **"telecommunication."**
Make sure to come back for our many
upcoming exciting adventures

See you later, alligator!

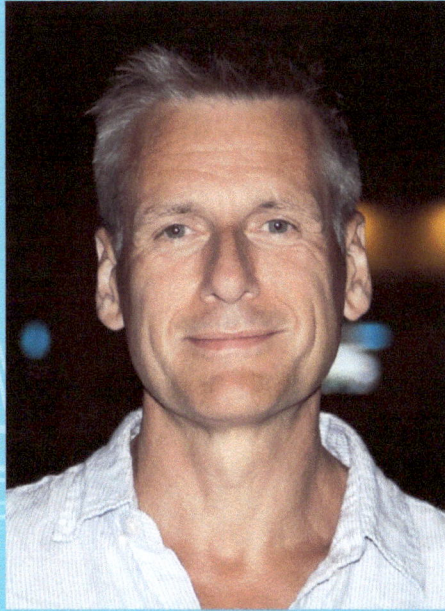

ABOUT THE AUTHOR

Gunnar Oddsson immigrated to the United States twenty-three years ago and for the last twenty years, he's been building and maintaining fiber-optic networks for all the major service providers on the East Coast.

After many years of long days and sleepless nights, Gunnar's focus is now on educating and informing the younger generation on how they communicate electronically and guiding children through the many hidden dangers of the electronic age.

With colorful and fun-loving cartoon characters, Gunnar's approach is simple but effective.

Gunnar is from Iceland, where books are still among the most popular Christmas gifts and where one in every ten people is likely to write and publish a book.

In Iceland they say, "Everybody walks with a book in his or her stomach."

Make sure to visit our website for fun and games
www.scuffleheads.com

www.ingramcontent.com/pod-product-compliance
Lightning Source LLC
Chambersburg PA
CBHW052045190326
41520CB00002BA/193